THE POETRY OF ERBIUM

The Poetry of Erbium

Walter the Educator

Silent King Books a WhichHead Imprint

Copyright © 2024 by Walter the Educator

All rights reserved. No part of this book may be reproduced in any manner whatsoever without written permission except in the case of brief quotations embodied in critical articles and reviews.

First Printing, 2024

Disclaimer
This book is a literary work; poems are not about specific persons, locations, situations, and/or circumstances unless mentioned in a historical context. This book is for entertainment and informational purposes only. The author and publisher offer this information without warranties expressed or implied. No matter the grounds, neither the author nor the publisher will be accountable for any losses, injuries, or other damages caused by the reader's use of this book. The use of this book acknowledges an understanding and acceptance of this disclaimer.

"Earning a degree in chemistry changed my life!"
- Walter the Educator

dedicated to all the chemistry lovers, like myself, across the world

CONTENTS

Dedication V

Why I Created This Book? 1

One - Symbol Of Curiosity 2

Two - Smallest Flame 4

Three - Igniting Hope 6

Four - Wonders We Can See 8

Five - Never Tire 10

Six - Every Space 12

Seven - Etched In Our Heart 14

Eight - Make Us Whole 16

Nine - Science Meets Art 18

Ten - Eternal Bliss 20

Eleven - Innovation And Change 22

Twelve - Paints Its Part 24

Thirteen - Testament To Science 26

Fourteen - Radiant Gem 28

Fifteen - Glory 30

Sixteen - Harmoniously 32

Seventeen - Soul 34

Eighteen - Unique Charm 36

Nineteen - Forever Shine 38

Twenty - Endless Grace 40

Twenty-One - Elegant Stance 42

Twenty-Two - Jewel 44

Twenty-Three - Nature's Craftsmanship . . . 46

Twenty-Four - Forever Captivating 48

Twenty-Five - Enchants 50

Twenty-Six - Profound 52

Twenty-Seven - Mystical Tale 54

Twenty-Eight - Maestro Of The Glow 56

Twenty-Nine - Unlocks The Door 58

Thirty - Magnetic Element 60

Thirty-One - Eternally Aligns 62

Thirty-Two - Minds And Hands 64

Thirty-Three - Rainbow Of Colors 66

Thirty-Four - Prism Of Colors 68

Thirty-Five - Erbium, The Element 70

About The Author 72

WHY I CREATED THIS BOOK?

Creating a poetry book about the chemical element Erbium serves multiple purposes. Firstly, it allows for a unique exploration of the element's properties, history, and significance. Poetry can capture the essence of Erbium in a way that scientific literature cannot, evoking emotions and painting vivid imagery. Secondly, it can make science more accessible and engaging to a wider audience. By combining scientific facts with artistic expression, this book can bridge the gap between the scientific and artistic communities. Lastly, it can inspire curiosity and spark interest in chemistry, encouraging readers to delve deeper into the world of elements and their fascinating characteristics.

ONE

SYMBOL OF CURIOSITY

In the depths of science's realm, I find Erbium's charm,
A chemical element, with secrets to disarm.
Its atomic number fifty-nine, a rare treasure indeed,
A lustrous metal, a beauty that few eyes can perceive.

Born from supernovae's fiery explosion,
Erbium dances in the cosmos with celestial devotion.
In the Earth's crust, it lies, hidden and serene,
A silent guardian, waiting to be seen.

Erbium, oh Erbium, your properties so profound,
In lasers and amplifiers, your brilliance is renowned.
With your magnetic allure, you capture the light,
Guiding us through darkness, a beacon shining bright.

Your energy levels, a symphony of hues,
From green to pink, a palette we can't refuse.

In the fiber optic cables, you transmit information,
Connecting hearts and minds, fostering communication.
But beyond your science, Erbium, lies a deeper tale,
Of unity and balance, where love and peace prevail.
For within each atom, a universe resides,
Where beauty and wonder forever coincides.
Erbium, a testament to nature's grand design,
A symbol of curiosity, a spark that forever shines.
In the vast expanse of the cosmos, you remind,
That even in the smallest things, miracles we can find.

TWO

SMALLEST FLAME

In the realm of elements, Erbium does abide,
A marvel of nature, where secrets do reside.
With atomic number fifty-nine, it claims its place,
A metal rare and precious, with elegance and grace.

From the heart of a dying star, Erbium was born,
An echo of cosmic forces, in brilliance it adorns.
In the depths of the Earth, it lies concealed,
A silent guardian, its mysteries unrevealed.

Erbium, oh Erbium, a jewel of the periodic chart,
Your magnetic properties, they captivate the heart.
With lasers and amplifiers, your power does unfold,
Guiding us through darkness, in stories yet untold.

A symphony of colors, Erbium, you display,
From shades of green to pink, a vibrant array.
In fibers of communication, you bear the light,
Connecting souls and minds, in an endless flight.

But beyond the science, Erbium, lies a hidden tale,
Of unity and harmony, where love and dreams prevail.
Within each atom, a universe does reside,
A reminder of the miracles, in every stride.

Erbium, a testament to nature's grand design,
A shimmering reminder, in the cosmic divine.
In the vast expanse of the universe, you proclaim,
That beauty can be found, even in the smallest flame.

THREE

IGNITING HOPE

    Erbium, a silent alchemist of the Earth's domain,
In the periodic table, you hold a mystic reign.
Number fifty-nine, a hidden gem of rare grace,
With secrets woven in your atomic embrace.
    From supernovae's fiery burst, you emerged,
A celestial creation, an element submerging.
In the depths of the Earth, your presence concealed,
A guardian of wonders, yet to be revealed.
    Erbium, the conductor of light's symphony,
With lasers and amplifiers, you dance in harmony.
Your magnetic allure, a mesmerizing sight,
Weaving tales of fascination, in the darkest night.
    Through the fiber-optic cables, you transmit,
Whispers of knowledge, connecting bit by bit.
A catalyst for communication, a bridge of connection,
Uniting distant souls, fostering affection.

Beyond the science, Erbium, your essence whispers,
Of unity and balance, of cosmic elixirs.
In each atom, a universe unfolds,
Where miracles reside, waiting to be beheld.
Erbium, a testament to nature's grand design,
A reminder that beauty lurks in the sublime.
In the vast expanse of creation's grand scope,
You unveil the wonders, igniting hope.

FOUR

WONDERS WE CAN SEE

Erbium, an enigma in the realm of elements,
A luminary force, where wonderments are sent.
With atomic number fifty-nine, you hold your ground,
A metal rare and precious, with mysteries profound.

Born from cosmic chaos, in celestial birth,
Erbium emerged, an embodiment of worth.
Hidden deep within the Earth's terrestrial frame,
A guardian of secrets, untouched by flame.

Erbium, the conductor of light's symphony,
In lasers and amplifiers, you find harmony.
Your magnetic prowess, a captivating spell,
Guiding us through darkness, where revelations dwell.

From hues of green to pink, your colors entwine,
A kaleidoscope of beauty, captivating and fine.

Through fiber-optic strands, you transmit the sublime,
Connecting hearts and minds in a cosmic rhyme.
    But beyond your science, Erbium, lies a deeper tale,
Of unity and balance, where harmony prevails.
Within each atom's core, a universe takes flight,
A reminder of the miracles, hidden in plain sight.
    Erbium, a testament to nature's grand design,
A beacon of curiosity, forever to shine.
In the grand tapestry of existence, you decree,
That in the smallest fragments, wonders we can see.

FIVE

NEVER TIRE

In the cosmic forge of celestial birth,
Erbium emerged, a treasure of great worth.
A metal enigma, hidden from sight,
In the Earth's embrace, a secret held tight.

Erbium, oh Erbium, with magnetic allure,
In lasers and amplifiers, your power pure.
From vibrant green to rosy hues,
A palette of elegance, a chromatic muse.

Fiber-optic whispers traverse the land,
Carrying knowledge with a luminous hand.
Erbium, the conductor of light's grand symphony,
Connecting souls, bridging the realms of harmony.

But beyond your science, Erbium, lies a tale untold,
Of unity and balance, a story to behold.
Within each atom's core, a cosmic dance,
Where miracles unravel in a cosmic expanse.

Erbium, a testament to nature's grand decree,
A reminder that beauty can be found, for all to see.
In the smallest fragments, wonders reside,
A glimpse into the universe, an eternal guide.
So let us marvel at Erbium's mystic grace,
A celestial gem, illuminating space.
In the tapestry of elements, you inspire,
A reminder of the wonders that never tire.

SIX

EVERY SPACE

Erbium, an enigma in the realm of science,
With your presence subtly concealed, a captivating alliance.
In lasers and optical fibers, your magic unfolds,
Guiding us through the darkness, where secrets are told.

A symphony of colors, Erbium, you unveil,
From vibrant greens to delicate pinks, a cosmic trail.
Through the depths of communication, you transmit,
Connecting hearts and minds, in a harmonious fit.

But beyond the boundaries of scientific lore,
Lies a tale of unity and wonder, forevermore.
Within each atom's core, a universe unfurls,
Where miracles reside, in the smallest of pearls.

Erbium, a testament to nature's grand design,
A beacon of curiosity, a spark that ignites the divine.

In the cosmic dance of elements, you shine,
A reminder that beauty lies in the sublime.
   So let us embrace Erbium's captivating allure,
An element of mystery, that forever endures.
In the tapestry of existence, you leave a trace,
A symbol of the wonders found in every space.

SEVEN

ETCHED IN OUR HEART

Erbium, a luminescent gem, rare and pure,
In the depths of science, your essence secure.
From the emerald green to the rosy hue,
Your colors paint a picture, vibrant and true.
Through fiber-optic veins, you gracefully glide,
Connecting distant souls, side by side.
But beyond the reach of scientific bounds,
A tale of unity and harmony resounds.
Within each atom's core, a universe unfolds,
Where mysteries dance, secrets yet untold.
Erbium, a testament to nature's artistry,
A symbol of curiosity, embracing diversity.
In the symphony of elements, you harmonize,
A reminder that beauty can be found in disguise.
So let us celebrate Erbium's radiant grace,

A catalyst of wonder, in this cosmic space.
In the grand tapestry of existence, you play your part,
A beacon of light, forever etched in our heart.

EIGHT

MAKE US WHOLE

Erbium, a luminescent treasure of the Earth,
A gem hidden in the periodic table's girth.
With atomic number sixty-eight, you reside,
A symbol of knowledge, impossible to hide.

In the realm of science, you hold your might,
With magnetic properties, shining so bright.
Through the fibers of technology, you flow,
Enabling communication, a powerful show.

But beyond your scientific prowess, Erbium dear,
There's a tale of wonder, let me make it clear.
Within your atomic heart, secrets unfurl,
Whispering stories of creation's precious swirl.

Erbium, a guardian of cosmic melodies,
A conductor of light, dancing with harmonies.
In the vast cosmos, you paint a celestial art,
Guiding stargazers, igniting the human heart.

So let us celebrate Erbium's captivating glow,
A gift from the universe, a treasure to bestow.
In the symphony of elements, you have a role,
A testament to the wonders that make us whole.

NINE

SCIENCE MEETS ART

In the realm of elements, Erbium you reside,
A luminescent jewel, with secrets to confide.
Within your atomic structure, a story unfolds,
Of mysteries and wonders, waiting to be told.

Erbium, a conductor of light's gentle sway,
You guide the photons, in an exquisite display.
Through optic fibers, you weave connections tight,
Uniting souls afar, with pure and radiant light.

But beyond your scientific allure, Erbium dear,
A deeper essence, a profound atmosphere.
In the depths of your core, a cosmic symphony,
Where atoms dance and resonate in harmony.

Erbium, a testament to balance and grace,
A beacon of serenity in the vast cosmic space.
In the tapestry of elements, you play your part,
A reminder of unity, engraved in every heart.

So let us revel in Erbium's ethereal charm,
Embracing the magic, letting our spirits disarm.
For within this element, a universe thrives,
Where science meets art, and wonder survives.

TEN

ETERNAL BLISS

Erbium, oh element of rare beauty,
A tapestry of wonders, intricate and free.
With atomic grace, you dazzle the eye,
A celestial jewel, lighting up the sky.

In your core, secrets of creation reside,
A symphony of particles, dancing with pride.
Erbium, conductor of the cosmic orchestra,
Guiding the universe with wisdom and grandeur.

Through the fibers of light, you transmit,
A bridge of connection, where souls unite.
Erbium, the messenger of thoughts untold,
Weaving tales of love, in stories yet unfold.

In the realm of elements, you stand apart,
A symbol of harmony, a work of art.
Your presence, a reminder of balance and peace,
A gentle caress, granting our souls release.

So let us cherish Erbium's ethereal glow,
A luminary force, that continues to grow.
In the vast expanse of the cosmic abyss,
Erbium, you enchant us with eternal bliss.

ELEVEN

INNOVATION AND CHANGE

Erbium, a gem within the periodic table's realm,
A radiant element, like a celestial helm.
In the heart of matter, your secrets reside,
Unveiling the wonders that science can't hide.

Amidst the cosmic dance, your atoms align,
With a magnetic allure, so beautifully divine.
Erbium, conductor of the electromagnetic show,
Guiding energy's flow, with a mesmerizing glow.

Through the fabric of time, you transmit and connect,
Like a cosmic messenger, your essence reflects.
Light and sound, in harmony you bind,
Creating a symphony of frequencies, one of a kind.

In the fiber-optic realm, your presence is grand,
Enabling information to travel across the land.

Erbium, a catalyst for innovation and change,
A building block for progress, in a world so strange.
   So let us celebrate Erbium's captivating grace,
A symbol of unity, in this vast cosmic space.
In the tapestry of elements, you hold a special place,
Erbium, a testament to the marvels of our human race.

TWELVE

PAINTS ITS PART

In the realm of elements, Erbium resides,
A luminary force with secrets it hides.
Within its atomic core, a cosmic dance,
Where wonders unfold in a mystical trance.

Erbium, a conductor of light and sound,
A catalyst of energy, profound and profound.
In the fibers of existence, you silently glide,
A bridge between worlds, where connections reside.

From laboratories to the depths of space,
Erbium's presence leaves a shimmering trace.
Its magnetic allure, a captivating spell,
Drawing us closer, under its enchanting spell.

Let us marvel at Erbium's ethereal glow,
A symphony of luminescence, a celestial show.
In the tapestry of elements, it takes its stand,
A reminder of nature's infinite command.

So raise a toast to Erbium's radiant might,
A symbol of ingenuity, shining bright.
In the vast cosmic canvas, it paints its part,
A beacon of wonder, forever in our heart.

THIRTEEN

TESTAMENT TO SCIENCE

In the realm of elements, Erbium dwells,
A hidden gem, where fascination swells.
With atomic grace, it captures the eye,
A lustrous presence, soaring high.

Erbium, a conductor of light's grand symphony,
Guiding photons with effortless harmony.
Through fiber-optic veins, you gracefully glide,
Connecting worlds, where communication thrives.

In the tapestry of elements, you hold a key,
Unlocking mysteries, for all to see.
A catalyst for innovation, a spark of inspiration,
Erbium, you fuel progress with determination.

In the depths of science, your secrets unfold,
Revealing the wonders, waiting to be told.

With magnetic allure, you draw us near,
Unraveling the universe, crystal clear.
   So let us celebrate Erbium's brilliant reign,
A luminary force, defying the mundane.
In the cosmic dance, you play your part,
A testament to science, a work of art.

FOURTEEN

RADIANT GEM

In the realm of elements, a jewel of rare worth,
Erbium, a treasure of the Earth's vast berth.
With atomic grace, you shimmer and gleam,
An enigma of beauty, a poet's dream.

In the tapestry of chemistry, your essence shines,
A symphony of electrons, in elegant lines.
Magnetic and vibrant, your energy sings,
Guiding us through the cosmos on celestial wings.

Erbium, the conductor of light's grand parade,
Through optical fibers, connections are made.
A messenger of knowledge, you gracefully impart,
Linking minds and hearts, across the world's chart.

In laboratories, you unveil secrets untold,
Revealing the wonders of the universe's mold.
A catalyst for discovery, a catalyst for change,
Erbium, you ignite the flame of the strange.

So let us cherish your presence, Erbium dear,
An element of wonder, both far and near.
In the vast cosmic dance, you hold your own,
A radiant gem, in the universe's throne.

FIFTEEN

GLORY

In the realm of elements, Erbium gleams,
A luminary presence within our dreams.
With atomic beauty, it captures the eye,
An ethereal gem in the cosmic sky.

Erbium, conductor of the electromagnetic symphony,
Harmonizing wavelengths in perfect unity.
Through optic fibers, it weaves a tapestry,
Connecting souls, transcending boundaries.

In the depths of science, its secrets unfurl,
Revealing its magnetic pull, a captivating swirl.
Erbium, a catalyst for technological might,
Igniting innovation, shining bright.

Within its core, a celestial dance,
Atoms swirling, in a cosmic trance.

A testament to nature's grand design,
Erbium, a jewel that continues to shine.
 Let us marvel at Erbium's radiant glow,
A beacon of knowledge, we've come to know.
In the vast expanse of the universe's story,
Erbium, you etch your mark in all of its glory.

SIXTEEN

HARMONIOUSLY

In the realm of elements, Erbium stands tall,
A symphony of atoms, mesmerizing us all.
With its magnetic allure, it pulls us near,
Revealing secrets of the universe, crystal clear.

A conductor of light, in fibers it dances,
Transmitting signals with elegant advances.
Through the vast expanse of space and time,
Erbium connects us, weaving a cosmic rhyme.

In laboratories, its mysteries unfold,
Unveiling the wonders that nature beholds.
A catalyst for progress, a spark of innovation,
Erbium fuels our quest for exploration.

With its radiant glow, it lights up the night,
Guiding us forward, with unwavering might.
A symbol of unity, in a world so vast,

Erbium reminds us that all things are connected, steadfast.

So let us celebrate this element divine,
Erbium, a jewel in the periodic line.
In the grand symphony of the atomic sea,
Erbium plays its part, harmoniously.

SEVENTEEN

SOUL

In the realm of elements, Erbium stands tall,
A luminary presence, captivating us all.
With a magnetic charm, it pulls us near,
Unveiling the mysteries that lie crystal clear.

Like a conductor leading a symphony grand,
Erbium orchestrates connections, hand in hand.
Through optic fibers, it carries the light,
Uniting distant souls, shining so bright.

In laboratories, its secrets are unveiled,
Revealing the wonders that nature has veiled.
A catalyst for progress, innovation's spark,
Erbium ignites the flame, leaving its mark.

A beacon of luminescence, it radiates,
Guiding us forward with its vibrant traits.
In the cosmic dance, it takes its place,
An element of beauty, full of grace.

So let us embrace Erbium's captivating allure,
A symbol of unity, steadfast and pure.
In the tapestry of elements, it plays its role,
Erbium, the embodiment of harmony and soul.

EIGHTEEN

UNIQUE CHARM

In the realm of elements, Erbium stands tall,
A conductor of marvels, captivating us all.
With its atomic dance, a symphony of grace,
Erbium weaves its magic in time and space.

In the fibers of light, it finds its home,
Connecting distant lands, no matter how far they roam.
Through the corridors of knowledge, its whispers resound,
Erbium, the messenger, profound.

In labs of exploration, its secrets unfurl,
Unveiling the mysteries of this intricate world.
A catalyst for discovery, a catalyst for awe,
Erbium, the enigma, defying every law.

With its radiant glow, it paints the night sky,
A celestial artist, captivating the eye.

A symbol of unity, amongst the cosmic array,
Erbium shines bright, guiding the way.
   So let us celebrate Erbium's unique charm,
A jewel in the periodic table, full of warm.
In the tapestry of elements, it claims its place,
Erbium, the element of wonder and grace.

NINETEEN

FOREVER SHINE

In the realm of elements, Erbium stands tall,
A luminary force, captivating us all.
Within its core, a symphony unfolds,
A dance of electrons, a story untold.
    With a vibrant glow, it paints the night sky,
A celestial beacon, shining up high.
Erbium's essence, a cosmic delight,
Guiding our spirits through the darkest of night.
    In laboratories, its secrets emerge,
Unlocking the mysteries, like a sacred surge.
A catalyst for progress, innovation's muse,
Erbium's presence, we cannot refuse.
    With magnetic allure, it pulls us near,
Drawing connections, crystal clear.

Erbium, the conductor of energy's flow,
Unveiling pathways where knowledge may grow.
    So let us celebrate Erbium's gleaming grace,
A symbol of curiosity, filling every space.
In the grand tapestry of the universe's design,
Erbium, a jewel that will forever shine.

TWENTY

ENDLESS GRACE

In the realm of elements, Erbium stands tall,
A luminescent marvel, captivating us all.
With atomic grace, it dances in the core,
Unveiling its secrets, forever to explore.

Through the fabric of science, it weaves its threads,
Connecting the dots, where knowledge spreads.
Erbium, the conductor of light's symphony,
Guiding us forward with a vibrant harmony.

In laboratories, its mysteries unfold,
Revealing its powers, like a tale untold.
A catalyst for progress, igniting invention,
Erbium fuels innovation with eternal ascension.

With its radiant glow, it paints the night sky,
A celestial brushstroke, pleasing to the eye.
Erbium, the jewel of the periodic chart,
A treasure of nature, a work of art.

Let us celebrate Erbium, the element divine,
A symbol of brilliance, forever to shine.
In the grand cosmic dance, it takes its place,
Erbium, a source of wonder and endless grace.

TWENTY-ONE

ELEGANT STANCE

In the realm of elements, Erbium finds its home,
A luminary force, uniquely its own.
With an atomic grace, it dances through the core,
Erbium, the essence that we deeply adore.

In laboratories, its secrets unfurl,
Unveiling the wonders of this precious pearl.
A catalyst for innovation, a spark so bright,
Erbium fuels our thirst for knowledge and light.

With its subtle glow, it illuminates the night,
A celestial guide, shining with might.
Erbium's symphony, a harmony so profound,
Uniting the universe, without a sound.

Let us celebrate Erbium's captivating appeal,
A symbol of curiosity, with a magnetic zeal.
In the tapestry of elements, it stands apart,
Erbium, the conductor of science and art.

So let our hearts embrace Erbium's grand embrace,
A gem of the periodic table, full of grace.
In the cosmic ballet, it takes an elegant stance,
Erbium, the element that enchants with its advance.

TWENTY-TWO

JEWEL

In the realm of elements, Erbium resides,
A luminary presence that forever abides.
With atomic grace, it dances in the core,
Revealing its wonders, a mystery to explore.

Erbium, the conductor of light's embrace,
Guiding our quest, with elegance and grace.
Through optic fibers, it weaves a connection,
Uniting distant souls, fostering affection.

In laboratories, its secrets unfold,
Unveiling marvels, untold stories untold.
A catalyst for progress, innovation's spark,
Erbium ignites the fire, igniting the dark.

With its radiance, it paints the night sky,
A celestial painter, captivating the eye.
Erbium, a beacon of luminescent art,
Illuminating our world, right from the start.

So let us celebrate Erbium's unique charm,
A jewel in the universe, a cosmic alarm.
In the tapestry of elements, it takes its place,
Erbium, the catalyst of wonder and grace.

TWENTY-THREE

NATURE'S CRAFTSMANSHIP

In the realm of atoms, a treasure resides,
Erbium, the element where magic abides.
With its atomic dance, a symphony unfolds,
A mystic force, its story untold.

In laboratories, its secrets unfurl,
Revealing wonders, a scientific whirl.
A catalyst for discovery, it takes flight,
Erbium, the spark that ignites insight.

With a subtle glow, it casts a gentle hue,
A celestial painter, creating something new.
Erbium's palette, a blend of colors rare,
A cosmic masterpiece beyond compare.

Let us celebrate Erbium's ethereal grace,
A symbol of curiosity, a captivating chase.

In the grand tapestry of the universe's design,
Erbium, a jewel that will forever shine.
   So let our minds marvel at Erbium's allure,
A testament to nature's craftsmanship pure.
In the symphony of elements, it claims its place,
Erbium, the element of wonder and embrace.

TWENTY-FOUR

FOREVER CAPTIVATING

Erbium, a jewel of the periodic table,
A hidden treasure, enigmatic and stable.
In the realms of science, it holds its might,
A fascinating element, shining so bright.

With a magnetic charm, it draws us near,
Unveiling mysteries, making them clear.
Erbium, the conductor of energy's flow,
Guiding us to pathways we yearn to know.

In laboratories, its secrets we unveil,
Unlocking wonders, like a captivating tale.
A catalyst for progress, innovation's muse,
Erbium, the spark that ignites and infuses.

With its luminescent glow, it lights the night,
A celestial presence, a guiding light.

Erbium, the beacon in the vast cosmic sea,
A symbol of curiosity, urging us to see.
   Let us celebrate Erbium's radiant grace,
A testament to science's endless chase.
In the grand symphony of elements, it plays,
Erbium, forever captivating in its ways.

TWENTY-FIVE

ENCHANTS

Erbium, a whisper in the realm of elements,
A silent force that quietly presents.
With atomic grace, it dances in the core,
Unveiling secrets, forevermore.

In laboratories, its wonders are revealed,
A catalyst for progress, power unconcealed.
Erbium, the conductor of light's symphony,
Guiding us forward with unwavering harmony.

With its luminescent glow, it paints the night,
A celestial brush, creating celestial delight.
Erbium, the jewel of the periodic chart,
A shimmering treasure, a work of art.

Let us celebrate Erbium, the element divine,
A symbol of brilliance, forever to shine.
In the cosmic dance, it claims its place,
Erbium, a source of wonder and grace.

So let our hearts embrace Erbium's embrace,
A beacon of knowledge, a celestial embrace.
In the tapestry of elements, it stands tall,
Erbium, the element that enchants us all.

TWENTY-SIX

PROFOUND

Erbium, the gem of atomic allure,
A luminescent essence, pure and demure.
Within its core, a magnetic grace,
A captivating element in the cosmic space.

With its gentle glow, it paints the night,
A celestial artist, weaving dreams of light.
Erbium's palette, a spectrum so rare,
A kaleidoscope of hues, beyond compare.

Let us celebrate Erbium's mystical charm,
A catalyst for curiosity, a mind's alarm.
In the symphony of elements, it takes its place,
Erbium, the conductor of nature's embrace.

In laboratories, its secrets unfold,
Unveiling marvels, untold stories untold.
A catalyst for progress, innovation's spark,
Erbium ignites the fire, illuminating the dark.

So let us marvel at Erbium's ethereal dance,
A symbol of wonder, in a cosmic expanse.
In the tapestry of creation, it holds its ground,
Erbium, a treasure of knowledge profound.

TWENTY-SEVEN

MYSTICAL TALE

Erbium, a whisper in the atomic choir,
A symphony of elements, where it inspires.
With elegance and grace, it takes its place,
A luminescent jewel, shining in cosmic space.

In laboratories, its secrets we unveil,
Unveiling wonders, like a mystical tale.
A catalyst for progress, innovation's muse,
Erbium, the spark that ignites and infuses.

With a vibrant glow, it paints the night,
A celestial brush, crafting celestial light.
Erbium, the artist of the cosmic show,
Creating wonders, for all to behold.

Let us celebrate Erbium's ethereal grace,
A symbol of curiosity, in this timeless chase.
In the grand tapestry of the universe's design,
Erbium, a jewel that will forever shine.

So let our minds wander through Erbium's realm,
Exploring its essence, like a sacred helm.
In the symphony of elements, it claims its place,
Erbium, the element of wonder and embrace.

TWENTY-EIGHT

MAESTRO OF THE GLOW

Erbium, a shimmering gem of the Earth,
A marvel of nature, of infinite worth.
In the depths of the periodic sea,
It stands apart, shining brilliantly.

With a gentle glow, it illuminates the way,
Guiding scientists in their quest to play.
Erbium, the catalyst of discovery's flame,
Unveiling mysteries, bringing knowledge to claim.

In laboratories, its secrets unfold,
Revealing wonders, stories yet untold.
A conductor of light, a maestro of the glow,
Erbium, the luminary, stealing the show.

Let us celebrate Erbium's ethereal light,
A symbol of curiosity, burning bright.

In the grand tapestry of elements, it thrives,
Erbium, the element that forever strives.
   So let our hearts be captivated by its charm,
A beacon of science, free from any harm.
In the symphony of nature's design,
Erbium, the element that will forever shine.

TWENTY-NINE

UNLOCKS THE DOOR

Erbium, the luminary of atomic grace,
A gem in the realm of the periodic space.
With its magnetic might, it pulls us near,
A force that whispers secrets in our ear.
Erbium, the enigma that captivates our minds,
Revealing mysteries, where curiosity binds.
In the laboratory's embrace, it unlocks the door,
Unveiling pathways we've never explored before.
A catalyst for innovation, it sparks the flame,
Erbium, the catalyst that fuels our aim.
Let us celebrate Erbium's spectral dance,
A symphony of colors, a cosmic romance.
In the grand tapestry of the universe's design,
Erbium, the luminary that continues to shine.
So let us marvel at Erbium's radiant glow,
A beacon of knowledge, forever aglow.

In the symphony of elements, it takes its place,
Erbium, the element that we embrace with grace.

THIRTY

MAGNETIC ELEMENT

Erbium, the subtle luminary of the periodic chart,
A silent protagonist, etching its mark.
In the realm of atoms, it holds a mystique,
A magnet for curiosity, modest and unique.
With its magnetic allure, it draws us near,
A force of attraction, both bold and clear.
Erbium's magnetic field, a captivating force,
Guiding us through uncharted course.
Let us celebrate Erbium's magnetic charm,
A conductor of electricity, a powerful arm.
In the tapestry of elements, it finds its role,
Erbium, the guardian of the electric soul.
In laboratories, its secrets are unveiled,
Revealing wonders, as the unknown is hailed.
A catalyst for innovation, it fuels the fire,
Erbium, the element that never tires.

So let us marvel at Erbium's magnetic might,
A symbol of strength, shining in the light.
In the symphony of elements, let it be heard,
Erbium, the magnetic element, forever stirred.

THIRTY-ONE

ETERNALLY ALIGNS

Erbium, a jewel in the periodic table's crown,
A luminary that never ceases to astound.
In laboratories, its properties unfold,
A realm of wonders, waiting to be told.
A catalyst for discovery, it sparks the flame,
Erbium, the element that ignites fame.
Let us celebrate Erbium's atomic grace,
A symbol of brilliance, lighting up the space.
In the grand tapestry of elements, it shines,
Erbium, the star that eternally aligns.
Its magnetic fields, a captivating force,
Guiding us through realms without remorse.
Erbium, the guardian of magnetic might,
Unveiling secrets, revealing the light.
So let us marvel at Erbium's cosmic dance,
A testament to science, an enigmatic trance.

In the symphony of elements, it finds its place,
Erbium, the element that leaves a lasting trace.
With every discovery, Erbium takes flight,
An element of wonder, woven in the fabric of night.
In the pursuit of knowledge, it leads the way,
Erbium, the catalyst, forever here to stay.

THIRTY-TWO

MINDS AND HANDS

In the realm of chemistry, a gem so rare,
Erbium, a name that fills the air.
With a luminescent gleam, it enchants,
A mystical element that forever enchants.
Erbium, the conductor of light's grand show,
Painting the skies with a celestial glow.
From the deepest cosmos to earthly plains,
Its radiance illuminates, leaving no remains.
Let us celebrate Erbium's magic untold,
A catalyst of wonder, a story yet unfold.
In the grand tapestry of elements, it weaves,
Erbium, the element that inspires and achieves.
In laboratories, its secrets doth unfold,
Unveiling mysteries, as new stories are told.
A spark for innovation, it lights the way,
Erbium, the guide that colors our day.

So let us marvel at Erbium's allure,
A symbol of curiosity that will endure.
In the symphony of elements, it stands,
Erbium, the element that moves minds and hands.

THIRTY-THREE

RAINBOW OF COLORS

In the realm of elements, Erbium holds its might,
A luminary in the dark, a beacon of light.
With atomic grace, it dances in the core,
Erbium, the element we cannot ignore.

Its magnetic prowess, a captivating force,
Guiding us through mysteries, staying on course.
Erbium, the guardian of magnetic allure,
Unveiling secrets, forever to endure.

Let us celebrate Erbium's spectral symphony,
A rainbow of colors, a celestial harmony.
In the grand tapestry of elements, it weaves,
Erbium, the element that forever achieves.

In laboratories, its wonders are revealed,
Unleashing discoveries with infinite yield.
A catalyst of innovation, it sparks the flame,
Erbium, the element that ignites the game.

So let us marvel at Erbium's enigmatic glow,
A symbol of curiosity, forever in the know.
In the symphony of elements, it takes its role,
Erbium, the element that touches the soul.

THIRTY-FOUR

PRISM OF COLORS

In the realm of elements, Erbium shines,
A luminary with secrets, so divine.
Its atomic dance, a captivating sight,
Erbium, the element that ignites the light.

In laboratories, its mysteries unfold,
Unveiling wonders, untold stories untold.
A catalyst for progress, it sparks the way,
Erbium, the element that leads the fray.

Let us marvel at Erbium's vibrant hue,
A prism of colors, so vivid and true.
In the symphony of elements, it plays its part,
Erbium, the element that touches the heart.

From the core of Earth to the vast expanse,
Erbium's presence leaves a cosmic trance.
With magnetic allure, it draws us near,
Erbium, the element that whispers in our ear.

So let us celebrate Erbium's wondrous grace,
A symbol of curiosity, in this cosmic chase.
In the tapestry of elements, it weaves its tale,
Erbium, the element that will forever prevail.

THIRTY-FIVE

ERBIUM, THE ELEMENT

A shimmering star, Erbium's grace unfurls,
In the realm of elements, its presence swirls.
A conductor of light, with a mystical gleam,
Erbium, the luminary of the periodic dream.
　Its atomic dance, a cosmic ballet,
Guiding the way, in an elegant display.
With magnetic allure, it pulls us near,
Erbium, the magnet of curiosity we revere.
　In laboratories, its wonders take flight,
Unveiling secrets, like stars in the night.
A catalyst for progress, it sparks innovation's fire,
Erbium, the element that never tires.
　So let us marvel at Erbium's radiant glow,
A beacon of knowledge, wherever we go.

In the symphony of elements, it claims its place,
Erbium, the element that leaves a lasting trace.
   With each discovery, Erbium's story is told,
A testament to science, in the mysteries it holds.
In the tapestry of nature, it weaves its art,
Erbium, the element that enchants the heart.

ABOUT THE AUTHOR

Walter the Educator is one of the pseudonyms for Walter Anderson. Formally educated in Chemistry, Business, and Education, he is an educator, an author, a diverse entrepreneur, and he is the son of a disabled war veteran. "Walter the Educator" shares his time between educating and creating. He holds interests and owns several creative projects that entertain, enlighten, enhance, and educate, hoping to inspire and motivate you.

Follow, find new works, and stay up to date with Walter the Educator™ at WaltertheEducator.com

www.ingramcontent.com/pod-product-compliance
Lightning Source LLC
LaVergne TN
LVHW020133080526
838201LV00117B/3731